Written by
Winde Bertran, R.N.

Illustrations By
Nafeesa Arshad

Disclaimer

This book is for educational purposes only and does not provide medical advice. It is not intended to diagnose, treat, cure, or prevent any condition. as Scientific understanding is continually evolving.

First edition

C.R.U.M.S. in L.E.D.R.
The Little Explorers of Discovery and Research™

The Endocannabinoid System:
AKA "ECS"
Is an important part of Human Biology;
Helping to keep the Body in Balance!

By
Winde Bertran, R.N.

*Note to Parents & Caregivers: This ECS Explorer story introduces early anatomy and body-system concepts through gentle metaphors and whimsical illustrations. All diagrams are simplified and non-graphic, supporting safe and age-appropriate science learning for ages 7-11, And Beyond!

For My Husband Conrad, who, without his support,
this wouldn't have come to fruition!!!
And, to My Children Daniel, Imani, and Conrad Jr.,
Who I Love with All My Heart!
My GrandChildren, Leo our music lover &
Camillo our computer techy.
And, as Grand Nana would say...My Lovies!
My Supportive Familial & Collective Community;
Professional and Entrepreneurial.
The "ECS" Medical Academia (which, as of this writing,
is only 9-13% of US medical students), and
Parental Community At Large (in hopes of sharing
decades of hidden human biology...)
As this is meant to increase
your Evidence Based knowledge of
The Endocannabinoid System!
And most Importantly, my sister Melanie!
Your Unconditional Love & Support has been a Blessing!!!

The "ECS" Explorer
From Pre-K to Pre-Med,
And Beyond!™

C.R.U.M.S. IN L.E.D.R.:
The Little Explorers of Discovery and Research
ECS
EXPLORER
CAPSULE
L.E.D.R
We are the C.R.U.M.S.:
A Sciency Bunch,
We Learn and Explore &
Solve things Before lunch!

Meet the CREW:

Con, Ros, Ultra, Mari, and Sati too,
With L.E.D.R., our robot who always pulls through!

Con - Circulatory Explorer
Ros - Respiratory Explorer
Ultra - Urinary Explorer
Mari - Muscular Explorer
Sati - Skeletal Explorer

We're the Little Explorers of Discovery and Research,
and our super-smart robot friend L.E.D.R. is ready
to take us on an adventure through
the body's Balance System known as
the Endocannabinoid System,
aka "ECS"!

Ana
L.E.D.R
L.E.D.R. begins explaining the
Endocannabinoid System aka
"ECS" to the C.R.U.M.S.

"Today," said L.E.D.R., "we've got a big quest":

"To visit the system that helps you feel best!"

Keys and Locks balance us;
to stay calm and light.

The ECS helps keeps it just right!

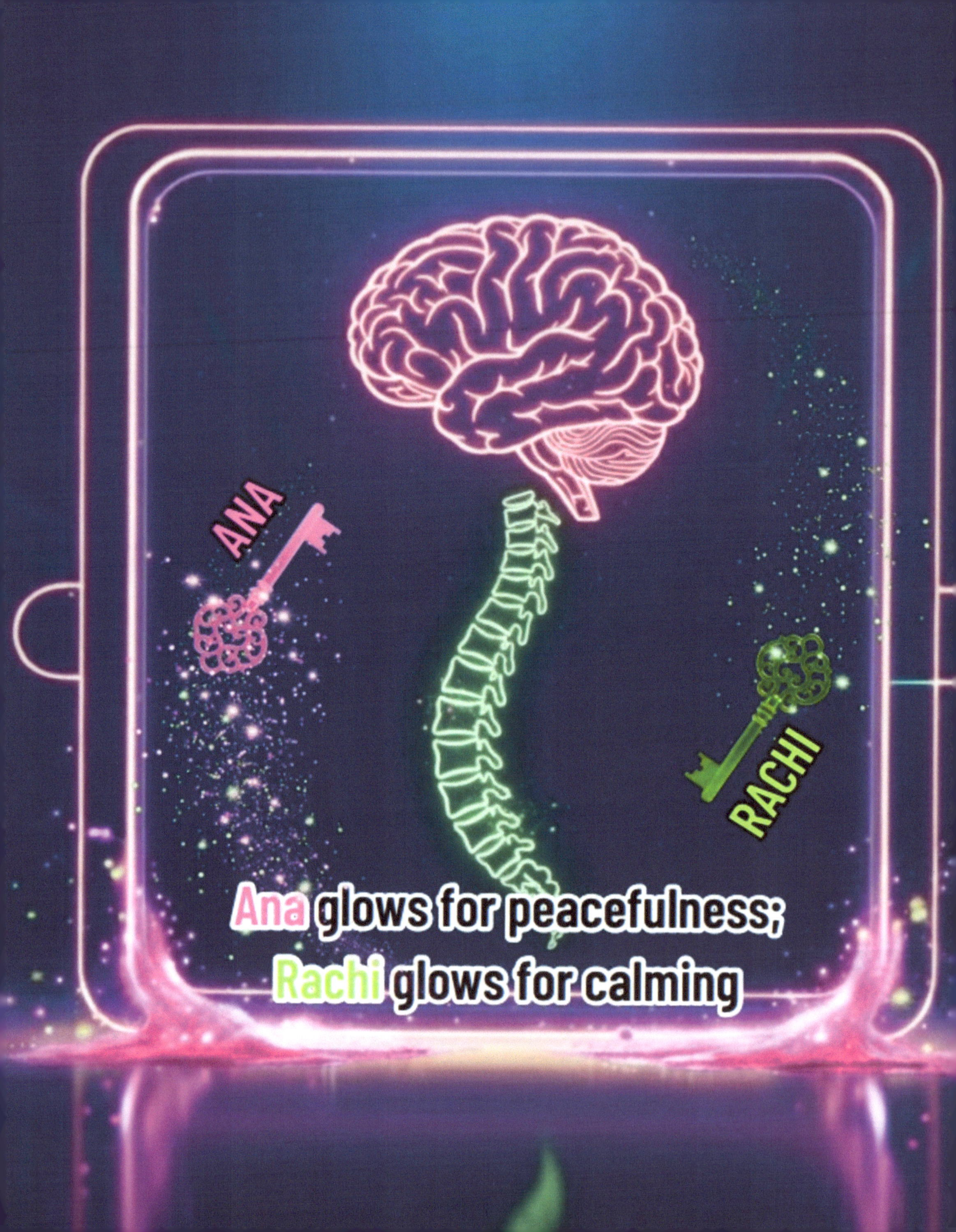
ANA
RACHI
Ana glows for peacefulness;
Rachi glows for calming

Two Key's appeared, full of sparkle and shine:

"Hi, I'm Anandamide; I help you feel fine!"

"I'm 2-AG, call me Rachi if you like"

We balance your body, like air in a bike!

Meet Receptors CB-1 and CB-2

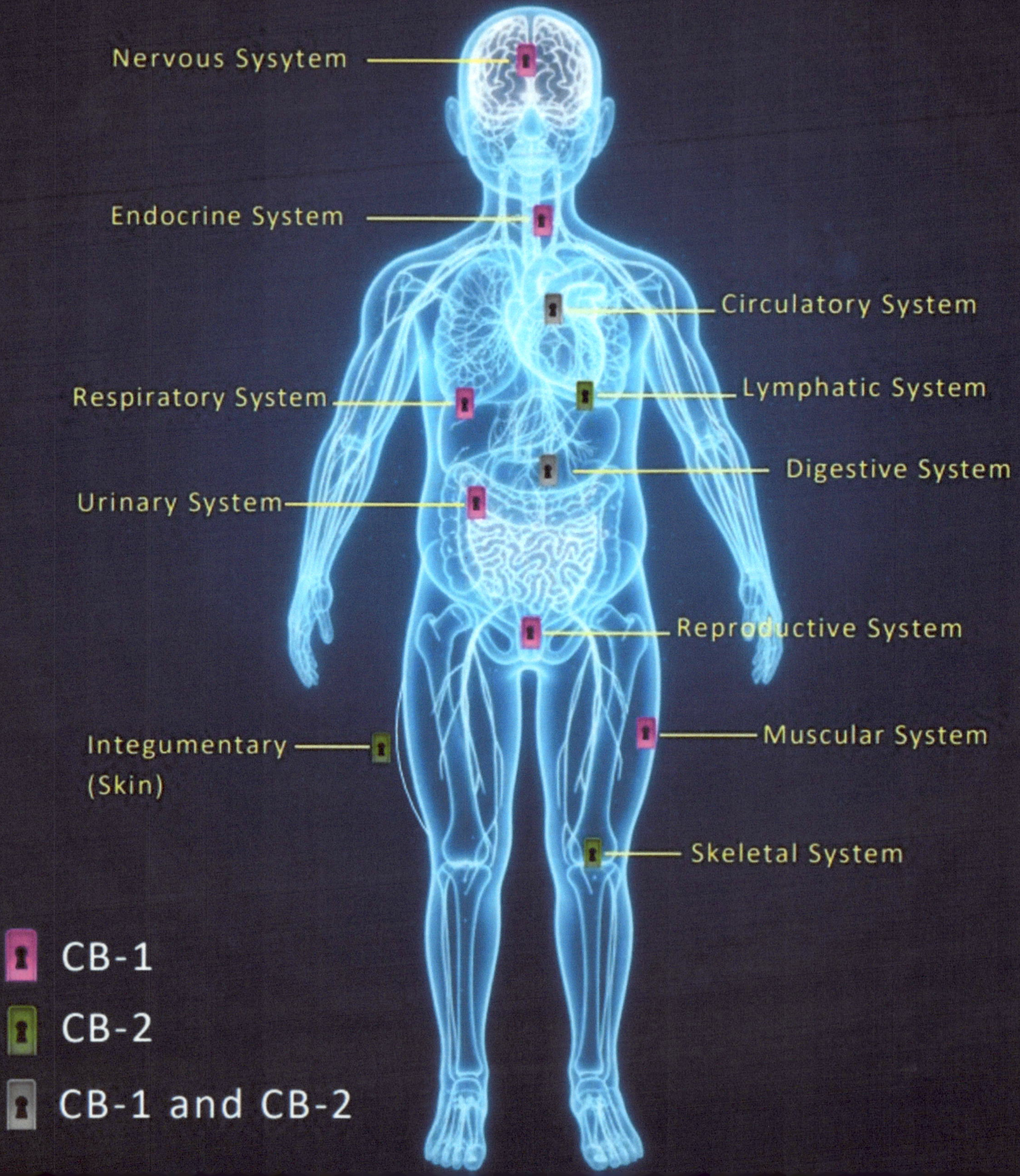

Keys travel to doors named CB-1 and CB-2
They help us know just what to do!

CB-1's in the brain, and the
nerves in your spine.

CB-2's within the body,
arms, toes, and legs that help you climb!

Enzyme Helpers

When Ana and Rachi are done with their chore,

Enzymes, like FAAH, appear and clean up the floor!

They break us down when we've finished our run.

The body stays balanced; their job is well done!

Circulatory System (C)

Thump, Thump, Thump, Thump, Thump, Thump!

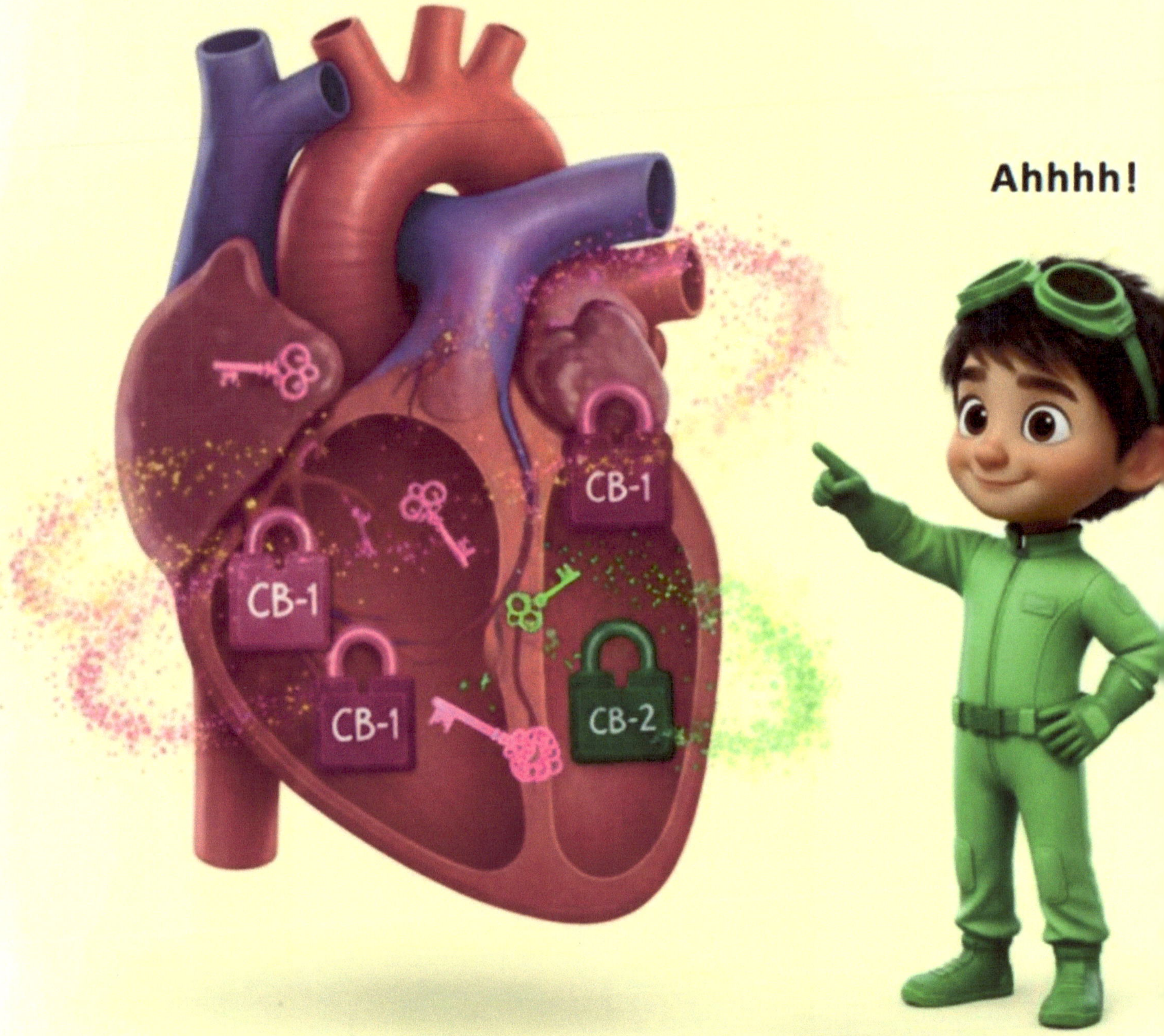

Ahhhh!

Thump, Thump…Thump, Thump!

Down Heart Highway, blood traffic was thick

The beat was too fast; it made Con feel sick.

The ECS helped his body find a calm beat.

Now blood flows smoothly,

from Con's head to his feet!

male reproductive system

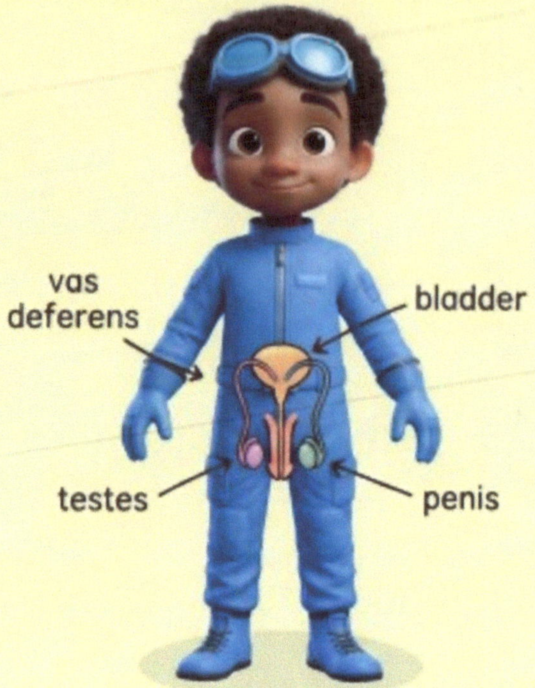

female reproductive system

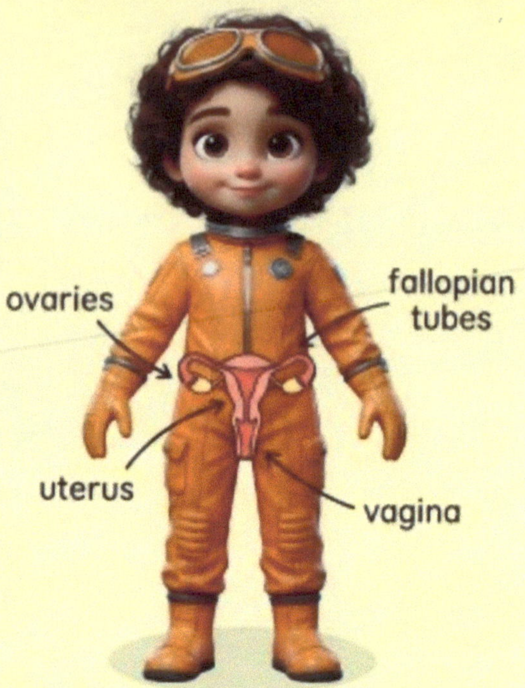

REPRODUCTIVE SYSTEM (R)

In Private-Parts Garden, the flowers were wild,

Growing too fast, and not quite mild.

CB-1 & CB-2 with Rachi gave growth some space,

So gardens can bloom, at a gentle pace.

URINARY SYSTEM (U)

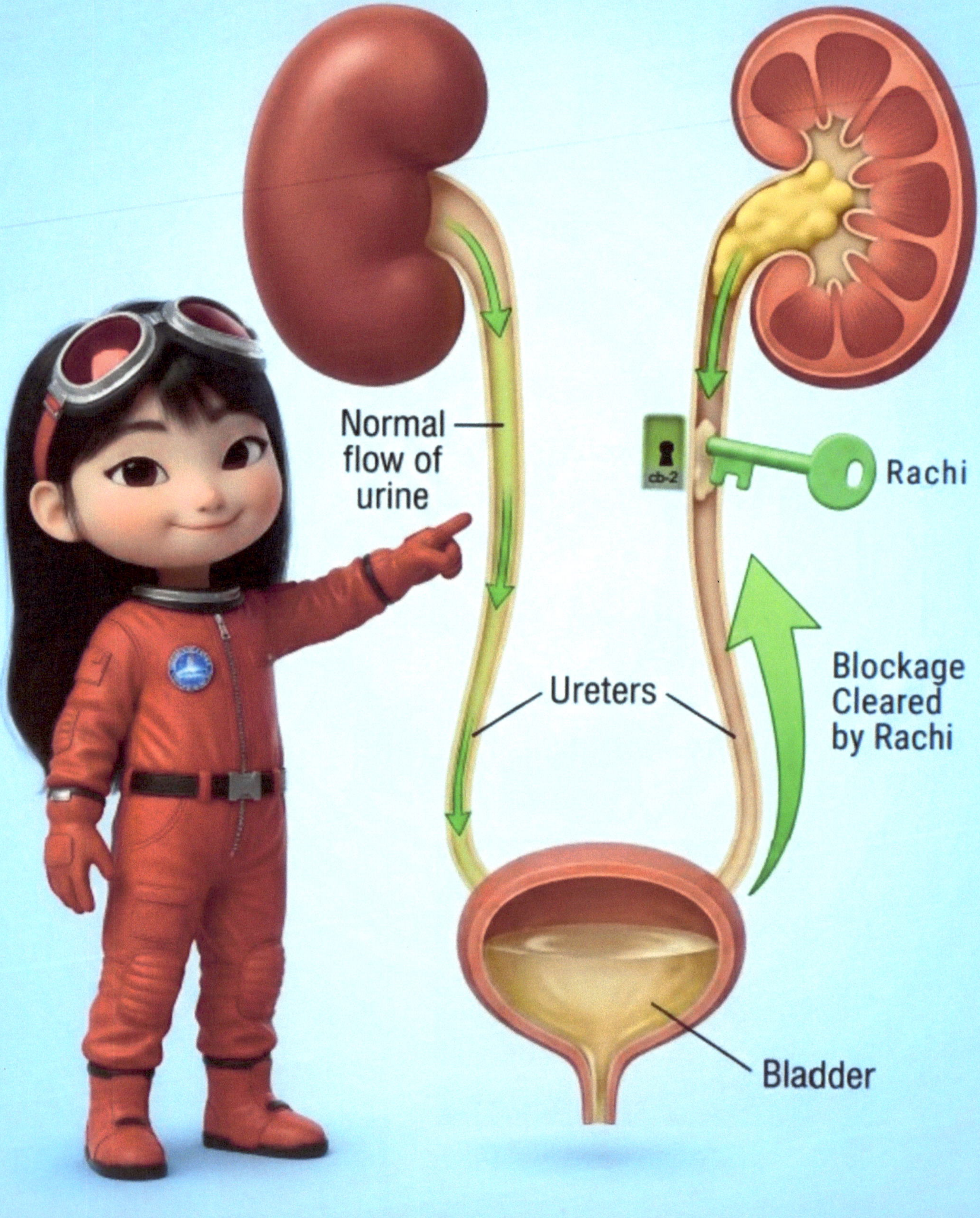

Down in the pipes of the Pee-Pee Canal,

Things were getting backed up..."oh no, pal!"

Ultra brought Rachi, whose key gave a wave.

And, along with CB-2, they helped the flow behave!

Muscular System (M)

During exercise, Mari's muscles got stiff as a board!

while stretching and moving, the pain she ignored.

Ana & Rachi whispered, "CB-2 knows what to do;

Relax now, muscles, because we believe in you!"

Skeletal System (S)

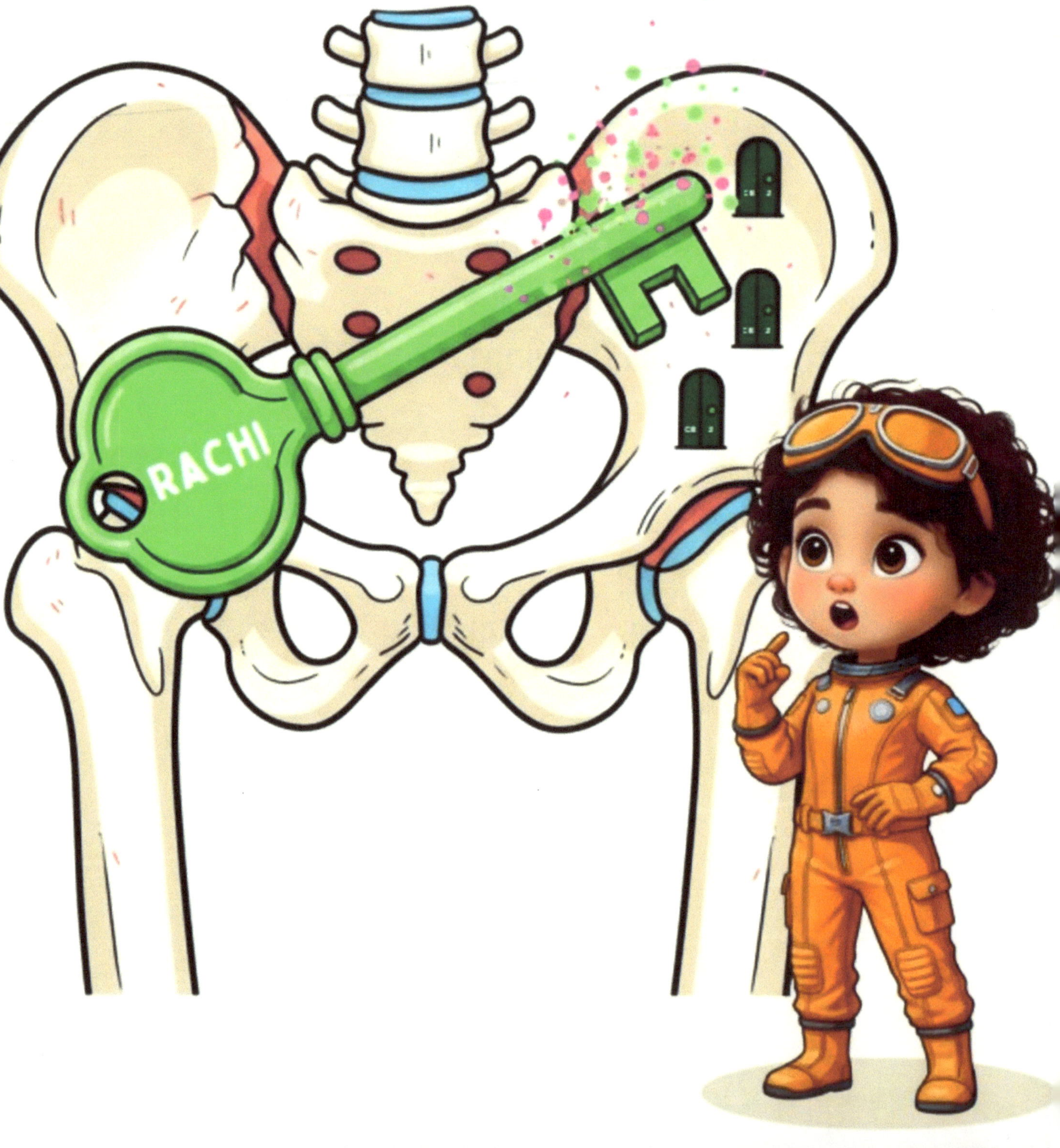

At Bone Bridge junction, Hips were clunky and creaky.

Sati said, "Strong bones shouldn't be this squeaky!"

Sati was amazed by the bone-builder crew,

learning that Rachi & CB-2 knew just what to do.

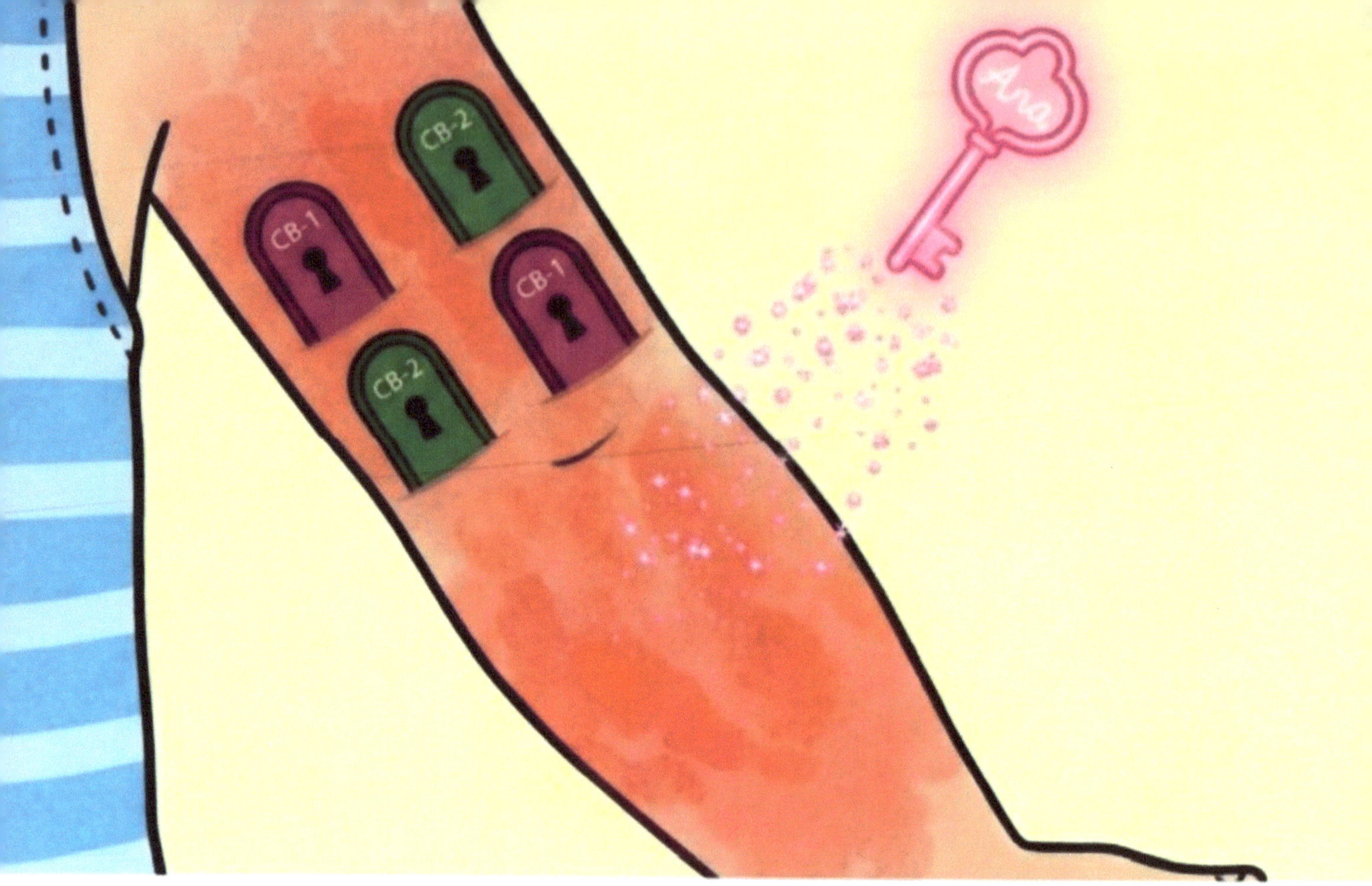

INTEGUMENTARY SYSTEM (I) AKA SKIN

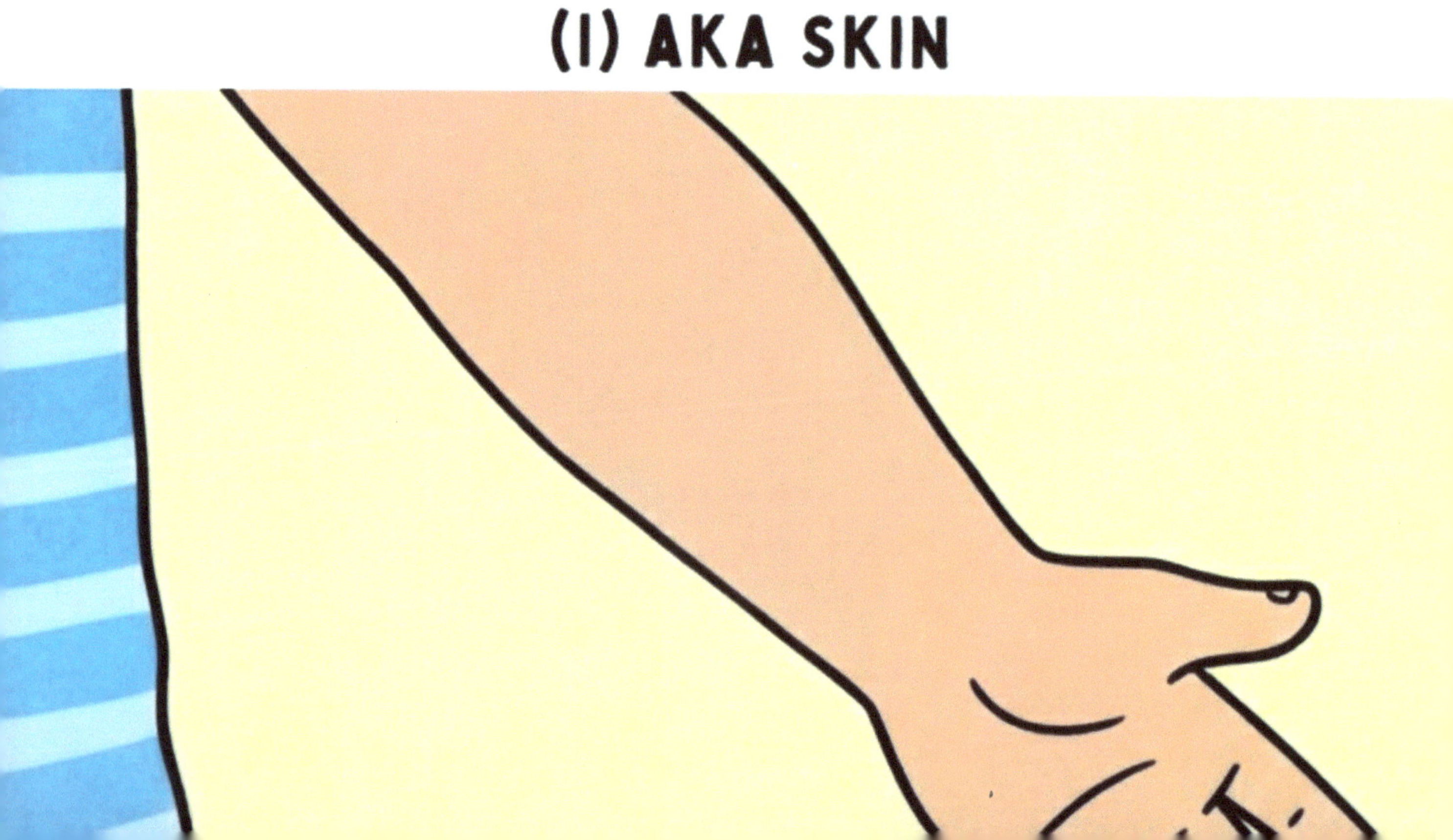

Skin City was itchy, red, and tight!

Scratching and flaking all day and all night.

CB-1 & CB-2 said "Please give them some rest!"

Ana cooled the skin, and now it feels best.

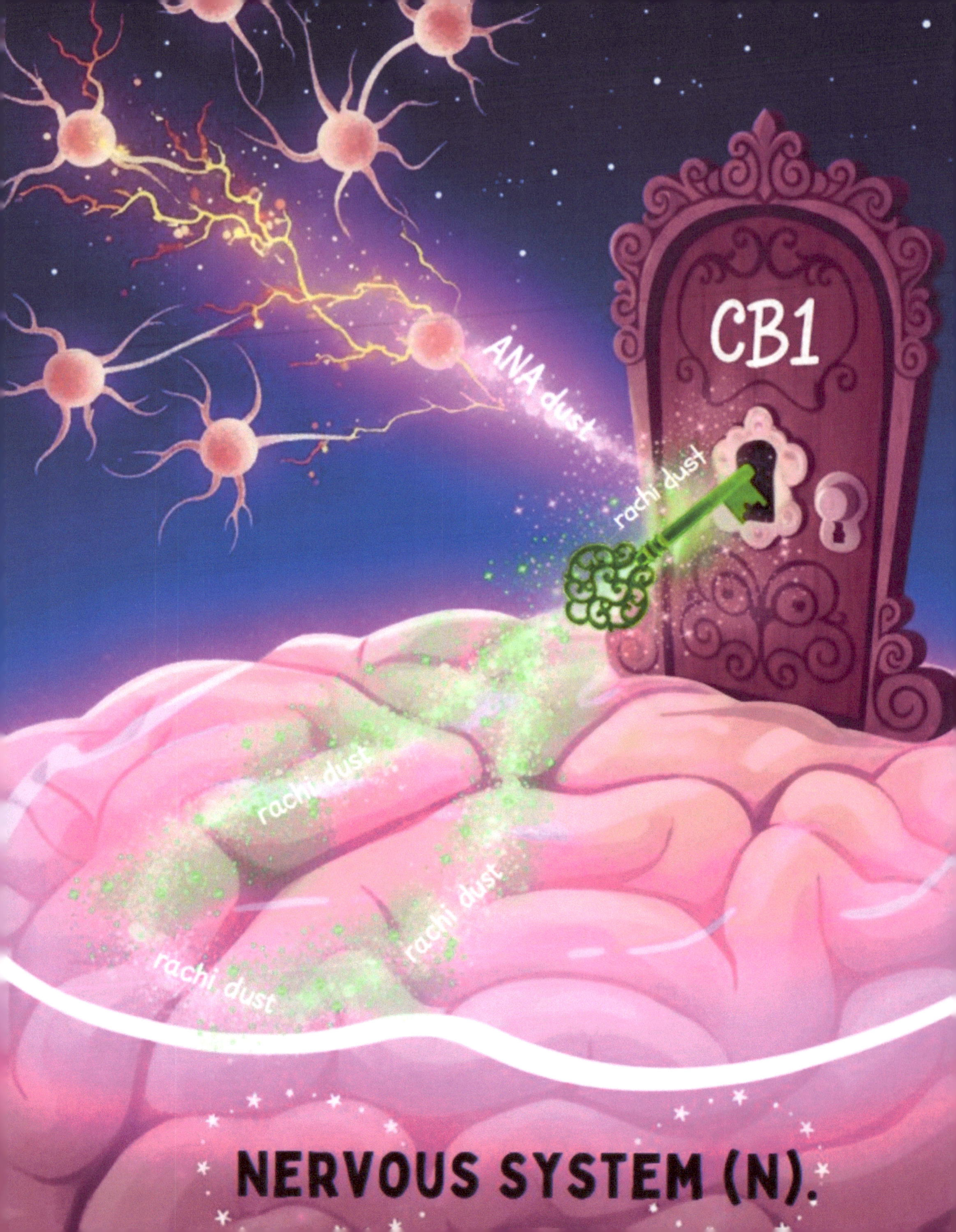

NERVOUS SYSTEM (N).

In Brainville sparks were zipping too fast;

Thoughts and emotions were flying right past!

Ana rang the bell on CB-1's door,

and said "Rachi will bring calm and smiles once more!"

LYMPHATIC SYSTEM (L)

Abnormal Throat

Normal Throat

Tonsil Park had Con's immune cells yelling,

"There's a little danger!" the T-cells were telling.

Rachi waved and said, CB-2's got your back.

"Let's take a break, and don't attack!"

ENDOCRINE SYSTEM (E)

Ros's Glandyland hormones were racing around.

Up, down, and flying off the ground!

Rachi calmed Cortisol, the stress hormone guy.

"Too much of you can make moods go awry!"

DIGESTIVE SYSTEM (D)

Gooey Gut Garden had food on the go:

The body wondered, 'is it moving fast or slow'?

After Stomach Signals gave Rachi & Ana a guide,

'We're Happy Gut Garden' the feelings replied.

Respiratory System (R)

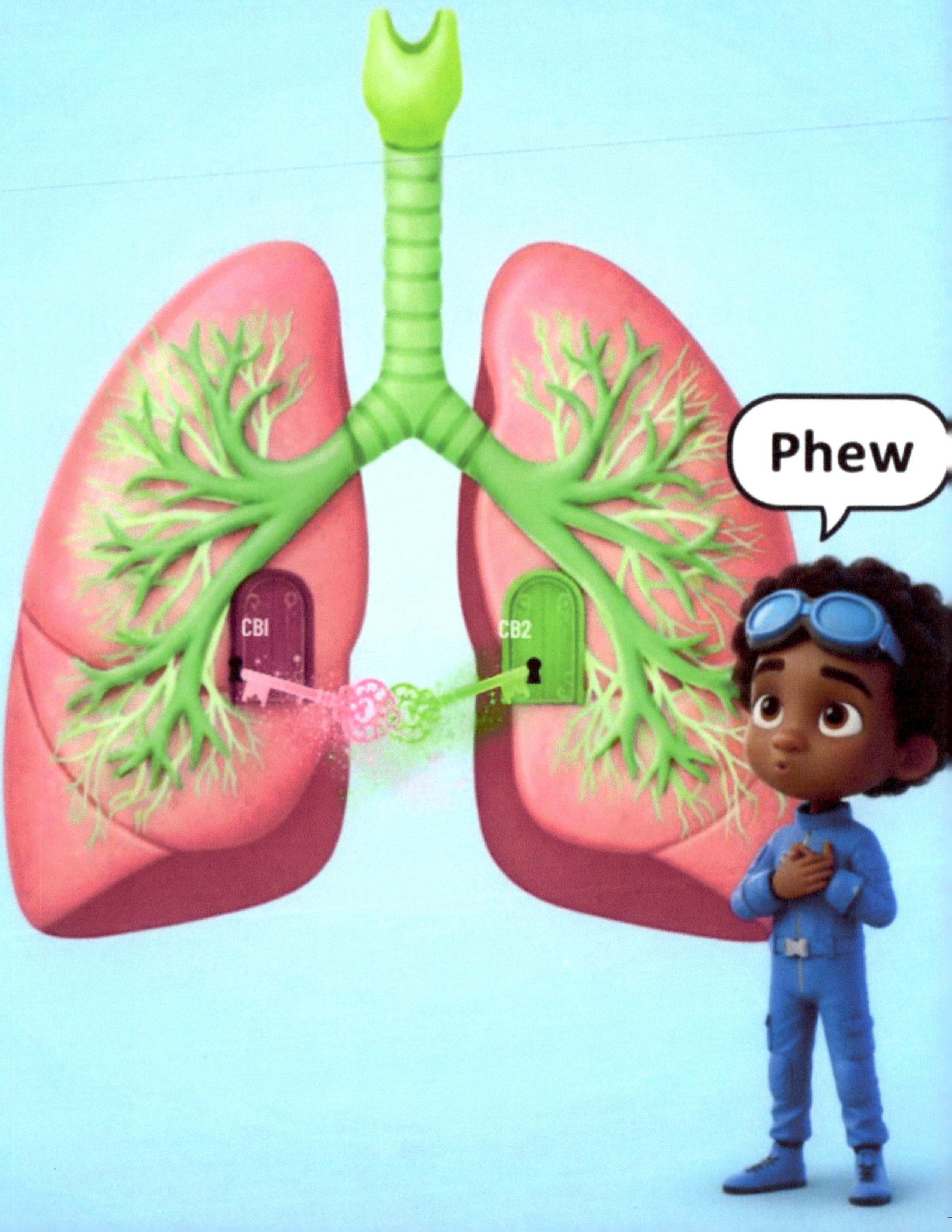

Lung Lagoon was wheezing with huffs & puffs;

The air was too tight, Ros's breathing was rough.

"CB-1 & CB-2 can help!" said Rachi with pride,

I'll calm the swelling, while Ana helps breaths glide.

MOOD CHECK-IN

Back in Brainville, a storm cloud spun.

Mari's Feelings were frowning, not having fun.

"Let's take a breath," said Ana & Rachi with care,

With a hug from Con, "we'll bring joy everywhere!"

ECS

just right
Rachi
Rachi
4
CB1
Ana

www.ingramcontent.com/pod-product-compliance
Lightning Source LLC
LaVergne TN
LVHW071239210926
848486LV00007B/308
9781970884005

ABOUT THE AUTHOR

Winde Bertran, R.N., is a Registered Nurse, educator, and entrepreneur with a lifelong passion for health, healing, and learning.

A wife, mother, grandmother, and advocate for curiosity at every age, she brings both science and heart to her work.

Through personal experience and professional practice, Winde believes there are no true "experts", only lifelong learners in an ever-evolving world.

Her children's books will blend playful storytelling with accessible science, inviting young readers and their families to explore the wonder of the human body together.

Did you know?

Scientists now believe the Endocannabinoid System (ECS) acts like a master balancing system, constantly adjusting mood, sleep, appetite, immunity, and stress...often before we even notice something is off.

***Think of it as your body's quiet problem-solver.**

Something amazing to think about:

The ECS is found in all vertebrates and works with the brain, immune system, and gut; showing that balance is a shared biological language across nature.

***Balance isn't just learned; it's built in.**

Curious minds note:

The ECS helps the body self-regulate: supporting balance, recovery, and resilience from the inside out.

*Bonus word: "FAAH" (Fatty Acid Amide Hydrolase) is a crucial "enzyme" in the body that acts as a "cleanup crew" for the Endocannabinoid System (ECS).

And, its primary role is to break down anandamide

Grown-Up's Corner

Grown-Ups, Did You Know?

The Endocannabinoid System helps regulate:

- **Mood**
- **Sleep**
- **Stress hormone (aka cortisol)**
- **Digestion**
- **Pain**
- **Immunity...and more!**

You can Support it with:

🌱 **Healthy food**
😴 **Restful sleep**
🏃 **Movement**
❤️ **Love and laughter**

Resources:

https://pmc.ncbi.nlm.nih.gov/articles/PMC5055136/

https://www.mdpi.com/1422-0067/26/3/1306

https://pmc.ncbi.nlm.nih.gov/articles/PMC3303140/

https://pmc.ncbi.nlm.nih.gov/articles/PMC11093854/

Bonus "ECS" Glossary - Big Science Words Made Easy!

Endocannabinoid System (ECS)

A special body system that helps all your other body systems stay balanced, like a coach that keeps the team running just right!

Ana (Anandamide) (Molecule)

A happy key your body makes to help you feel calm and safe.
Ana is like a cozy hug for your brain and tummy!

Rachi (2-AG) (Molecule)

A busy key that helps your body move, grow, and stay balanced.
Rachi helps your muscles, bones, and tummy, too!

CB-1 (Receptor)

Tiny doors in your brain and nerves.
Ana and Rachi knock when your brain needs help.

CB-2 (Receptor)

Tiny doors in your body Rachi visits, when your body needs to heal, calm down or attack germs.

Receptors

Special spots (doors) on your cells that receive messages; like doorbells!
Ana and Rachi press them to tell your body what to do.

Enzymes (FAAH)

The clean-up crew.
They tidy up the leftover's of Ana and Rachi when their job is done.

Cortisol

Is a stress signal.
A little is okay, but too much is not. Rachi helps slow it down.

Balance

When everything feels just right; not too fast, not too slow!

Bonus Word: Hypothalmus

A tiny control helper in your brain that helps your body know when to eat, sleep, feel warm or cool, and stay balanced;
like a lil boss making sure everything feels just right!

♫ Repeat After Me! ♫

Ana= Calm and peace we see!

Rachi = Helps us move and be!

CB1 = Brain and mood, that's right!

CB2 = Body's strength and healing light!

Enzymes = Clean when the job is done;

ECS balance is so much fun!

L.E.D.R
We are the C.R.U.M.S.:
A Sciency Bunch,
We Learn and Explore &
Solve things Before lunch!

They zoomed back out of the "ECS" core,
Feeling stronger and wiser than ever before.

Ana and Rachi gave a flying dance,
"Thanks for the balance;
We loved this chance!"

RETURN TO BASE

The crew gave high-fives and joyful hugs.
The body was balanced with no more bugs!

L.E.D.R. beamed, "You've learned so much;
From systems and science, to balance and such!"

ECS
EXPLORER
CAPSULE
MISSION COMPLETE!
LEDR
ULTRA
MARI
ROS
CON
SATI

In Dreamy Den with stars above,

Ultra said, "Without rest, we can't feel love!"

Rachi & Ana helped CB-1 feel cozy and deep,

And soon everyone settled into peaceful sleep.

SLEEP AND REST (BRAIN)
CB-1
Rachi
Ana

In Hypothalmus Cave, things weren't quite right:
Left side was freezing, Right side was bright!

"ECS" nudged the temperature control,
Ana warmed, Rachi cooled, and CB-1 took hold.

"Yeah!" said Sati, warm & cool found middle ground,
Now comfy balance was back around!